KT-450-074

GUIDES

PLUMBING

REAL LIFE GUIDES

Practical guides for practical people

In this increasingly sophisticated world the need for manually skilled people to build our homes, cut our hair, fix our boilers and make our cars go is greater than ever. As things progress, so the level of training and competence required of our skilled manual workers increases.

In this series of career guides from Trotman, we look in detail at what it takes to train for, get into and be successful at a wide spectrum of practical careers. *Real Life Guides* aim to inform and inspire young people and adults alike by providing comprehensive yet hard-hitting and often blunt information about what it takes to succeed in these areas.

The other titles in the series are:

Real Life Guide: The Armed Forces
Real Life Guide: The Beauty Industry
Real Life Guide: Care
Real Life Guide: Carpentry and Cabinet-Making
Real Life Guide: Catering
Real Life Guide: Construction
Real Life Guide: Distribution and Logistics
Real Life Guide: Electrician
Real Life Guide: The Fire Service
Real Life Guide: Hairdressing
Real Life Guide: The Motor Industry
Real Life Guide: The Police Force
Real Life Guide: Retailing
Real Life Guide: Transport
Real Life Guide: Working Outdoors
Real Life Guide: Working with Animals and Wildlife
Real Life Guide: Working with Young People

trotman

Real Life

GUIDES

PLUMBING

Dee Pilgrim
Second edition

Real Life Guide to Plumbing
This second edition published in 2007 by Trotman an imprint of
Crimson Publishing, Westminster House, Kew Road,
Richmond, Surrey TW9 2ND
www.crimsonpublishing.co.uk

© Trotman 2007

First edition written by Mike Hobbs, published 2004 by
Trotman and Co Ltd
Reprinted 2005

Editorial and Publishing Team
Author Dee Pilgrim
Editorial Mina Patria, Publishing Director; Jo Jacomb,
Managing Editor; Ian Turner, Production Editor
Production John O'Toole, Operations Manager
Sales and Marketing Catherine Slinn, Marketing Executive
Advertising Sarah Talbot, Head of Advertising

Designed by XAB

British Library Cataloguing in Publications Data
A catalogue record for this book is available from the British
Library

ISBN 978 1 84455 118 7

Typeset by Photoprint, Torquay
Printed and bound in Great Britain by
Creative Print & Design, Wales

Real
Life

GUIDES

CONTENTS

About the author

Dee Pilgrim completed the pre-entry, periodical journalism course at the London College of Printing before working on a variety of music and women's titles. She has written numerous articles and interviews for *Company*, *Cosmopolitan*, *New Woman*, *Woman's Journal* and *Weight Watchers* magazines. For many years she covered new output by singer/songwriters for *Top* magazine, which was distributed via Tower Records stores, and during this period interviewed the likes of Tori Amos, Tom Robinson and Joan Armatrading. As a freelancer for Independent Magazines she concentrated on celebrity interviews and film, theatre and restaurant reviews for magazines such as *Ms London*, *Girl About Town*, *LAM* and *Nine to Five*, and in her capacity as a critic she has appeared on both radio and television. She is currently the film reviewer for *Now* magazine and has written a number of titles for Trotman Publishing. When not attending film screenings she is active in the Critics' Circle and is the secretary for its film section.

Acknowledgements

Many people helped me with the second edition of the *Real Life Guide to Plumbing* but I would like especially to thank the following:

- Tony Beaumont
- Jenny Eltringham
- Danny Davis.

Dee Pilgrim
January 2007

Foreword

Everyone looks for a career that requires them to be highly skilled, customer focused and energetic. But not many careers combine all of that with working with tools, technical processes and with modern technology. The plumbing industry offers a broad range of opportunities and City & Guilds makes sure all our plumbing qualifications meet industry requirements.

Generally young people enter the plumbing industry on an Apprenticeship scheme. Apprenticeships are work-based training programmes where you can get hands-on experience in developing a career in plumbing.

There are various City & Guilds plumbing qualifications available depending on your previous skills and knowledge. However, if you have no previous experience or knowledge, generally you would need to start with the Certificate in Basic Plumbing Studies. After that the training possibilities are endless. There are several assessment methods to suit all learning styles, ranging from up-to-the-minute e-learning that can be combined with practical on the job work-placed assessment (NVQs) or college-based off-job learning, training and assessment/assignment work (VRQs).

However, if plumbing doesn't interest you then there are alternative qualifications available in gas and heating and ventilation too.

But watch out! Due to the increase in demand for training for plumbing and allied trades, a number of organisations have set up and sell their own plumbing training courses in plumbing skills and knowledge, implying qualified status, employment or franchise businesses. Generally the training will take place over short period of time, making it more attractive to mature student who wish to change their career. This kind of training can be very expensive and **may not** lead to a plumbing sector/industry recognised qualification.

City & Guilds are delighted to be part of the Trotman *Real Life Guides* series to help raise your awareness of these vocational qualifications. If an opportunity of involvement in plumbing is for you, City & Guilds has qualifications which will support development throughout your career, helping you to achieve excellence and quality in whatever field you choose.

Chris Johnson
Plumbing Manager
City & Guilds

Introduction

It's every home owner's worst nightmare: arriving back at your flat or house to find a giant puddle under the main radiator; or discovering your central heating has decided to die in the middle of winter; or worse still flushing the toilet only for the contents of the toilet bowl to back up and empty themselves on your bathroom floor. In each of these examples there is only one logical course of action open to you – flick through your local telephone directory and find an affordable and reliable emergency plumber. However, that may not be as easy as it sounds, for although there may be many plumbers listed, how can you tell which have been properly trained and will not overcharge you?

It's a thorny problem and not one that's about to disappear overnight, for the plumbing industry has been going through a volatile period recently. As little as two years ago the CITB-ConstructionSkills, a partner in ConstructionSkills, the Sectors Skills Council for construction, was predicting that over the next five years there would be a shortfall of 30,000 plumbers across the UK, mainly because there weren't enough young people coming through training courses or apprenticeship schemes. Sniffing a good story, the British press started printing articles about how demand for good plumbers was outstripping supply and how this had led to some plumbers earning over £70,000 a year. The upshot was that accredited courses in plumbing across the UK suddenly became massively oversubscribed and not always by the best candidates, while a new breed of 'rogue trainer' (see the box on page 49) sprang up charging thousands of pounds for courses offering no formal qualifications whatsoever. Add to this the expansion of the EU and an influx of Eastern European plumbers with little or no knowledge of British building regulations and the result has been that there now seem to be plenty of plumbers out there, it's just that many of them are unskilled, badly trained and inexperienced.

'It's not that there is a shortage of labour, it's still a skills shortage,' explains Danny Davis, the Technical Officer for the Institute of Plumbing and Heating Engineering (IPHE). 'There are a lot of people out there who are calling themselves plumbers, including many from Eastern Europe, but they don't necessarily have the appropriate qualifications. It's a real problem because what happens is properly trained plumbers are then having to come in and do remedial work when things go wrong.' This is backed up by a recent survey the IPHE conducted among its members. Many believed they had lost work by being undercut by Eastern European plumbers and the majority said a high percentage of the work they were carrying out was remedial, caused by incompetent plumbers who had broken water and building regulations.

It's not that there is a shortage of labour, it's still a skills shortage

That may make it sound like now is not a good time to be thinking about plumbing as a career, but in fact, the opposite is true. The industry is crying out for more capable people to come through the system and be properly trained on officially accredited courses. Tony Beaumont is the Curriculum Manager for Construction at South Thames College where they have only been offering Technical Certificates in plumbing, heating and ventilation for the last three years. There are currently 240 full-time and part-time students studying for the Certificates but Tony says they could put another 240 places on tomorrow and they would all be filled. 'We are always oversubscribed and during enrolment the plumbing courses are filled within two days,' he says. 'But no matter how many people we train there is always a shortage of good plumbers.'

This is especially true as the nature of the work is changing rapidly. New technology and legislation about energy conservation and environmental issues mean modern plumbers must be better trained and more knowledgeable than ever before. Both Danny Davis and Tony Beaumont believe the problems the industry has recently been facing could be resolved if all plumbers were required by law to be properly qualified. Although this is something the IPHE has been pushing for many years, there is no current legislation going through Parliament, but things are changing. 'We are currently in the middle of new requirements coming in,' says Danny Davis. 'Within the next couple of years you are going to be required to be "competent", meaning if you want to do plumbing and heating your work must be signed off and certified, and the certificate can only be issued by a "competent" person. That's why Competent Plumber Schemes (CPS) will be introduced and we certainly see the need for competent plumbers growing, especially with the amount of new building, such as the Olympic construction site, now being undertaken in this country. There's a lot of promise out there for the industry.'

The problems the industry has recently been facing could be resolved if all plumbers were required by law to be properly qualified

So, you may not be able to earn £70,000 as a plumber, but there's still plenty of work out there and with all the new developments in the industry this can be a hugely satisfying job for life.

DID YOU KNOW?

People with qualifications at National Vocational Qualification (NVQ) Level 3 (or SVQ Level 3 in Scotland) earn on average 25 per cent more each year than those of the same age and experience with no qualifications.

This guide will show you the types of skills and attitudes you will need to make a career in plumbing a success. There are plenty of stories and advice from people in the profession to let you know what you should expect. Obviously you won't learn everything from a book, but it will give you a flavour of the different jobs in the industry, its good and bad points, and help you decide whether becoming a plumber is really for you. It will offer information on what courses and qualifications you will need in order to progress and also examines the broader opportunities in plumbing. Whether you want to work by yourself and for yourself, be one of a team, or oversee your own team in a management capacity, plumbing can be a very fulfilling way of earning your living. This book can help you make an informed decision, and remember, somewhere in the United Kingdom at this very minute someone's about to phone for a plumber!

CHRISTOPHER HANLON

Success story

MANAGING DIRECTOR

Christopher Hanlon started in the plumbing industry several years ago and has now worked his way up to be Managing Director of his own company in Glasgow, Scotland.

How did you get your first job – and what was the training like?
I had a very enjoyable apprenticeship with the company I served my time with. It was a company that carried out work similar to the one I now run, in other words within households. I found it both enjoyable and educational from the point of view that we learned various different trades within the plumbing industry.

What do you do now?
I run a Glasgow-based plumbing and heating company that serves the domestic and household repair market.

How long have you been in business for yourself?
I started the business over 15 years ago and we've been constantly expanding ever since.

A young plumber should be ambitious, have a solid layer of general knowledge but be willing to learn

How many people do you employ?
We now employ between 30 and 35 staff, mainly depending
on the time of year. For a number of years we have also
been offering a 24-hour emergency service, but recently we
have set up a separate specialist company to service the
emergency market.

*Have you any advice for school leavers wishing to become
plumbers?*
In my experience, a young plumber should have a few
simple, overriding attributes and qualities. He or she should
be ambitious, and want to do well, and have a solid layer of
general knowledge but be willing to learn.

To be successful, the apprentice must be enthusiastic
and hard-working, dress smartly and always be
respectful of others

*Can you identify any qualities that make people more likely to
succeed?*
To be successful, the apprentice must be enthusiastic and
hard-working, dress smartly and always be respectful of
others, which also means being polite and trustworthy, in
order to gain the customer's confidence.

What's the story?

Many people only ever get to see a plumber at work when they have an emergency, such as a pipe bursting, or when a major plumbed-in fixture, such as a boiler, needs to be replaced. But what, actually, does a plumber do? Basically, plumbing is all about water, pipes and drainage. A plumber installs, repairs and replaces everything from kitchens and bathrooms to radiators, washing machines and dishwashers. The main areas are sanitation systems (such as toilets and sinks), heating systems (these may be fuelled by oil, gas or a form of solid fuel such as coal), outside guttering work, underground pipework, and even weather proofing on roofs. Some plumbers work in private houses, others work on building sites, still others work as maintenance plumbers for large companies. Whether the working environment is residential, commercial or industrial, you'll find that only a qualified plumber has the skills required to install and service the massive variety of plumbing and heating equipment available.

TECHNOLOGICAL ADVANCES/RENEWABLES

There's already more than enough work for everyone, and with the constant advances in plumbing and heating technology there are bound to be new employment opportunities for plumbers well into the future. 'The biggest trend for the future will definitely be energy and water efficiency, including renewables,' says Danny Davis. 'In 2005 new legislation came in saying everybody now has to have a condensing or high efficiency boiler. This year there are going to be changes to the way kitchens and bathrooms are installed and much of this is to do with the government wanting better energy and water efficient products to be used. Energy efficiency has been on the agenda for some time, but now water efficiency is becoming more important due to the recent droughts. Over the next three or four years renewables such as solar energy, ground source heat pumps and some form of rainwater collection connected to your water system will become more and more prevalent.' Tony Beaumont agrees with him and adds: 'These days, a lot of new builds only get the go ahead for the contract if they address environmental issues such as energy and water conservation, whether that's recycling rainwater or water efficient taps and showers. Now some of the larger plumbing firms are concentrating solely on the installation of solar panels.'

Although different plumbers do different things, the essence of the job remains the same; describing the job is therefore fairly straightforward and clear-cut.

The main skill you'll need is to be able to **cut pipework**, because all water flows in and out of buildings via pipes. You'll have to do this carefully and accurately, ensuring that all pieces and joints fit together.

The other especially useful skill you'll need to acquire in cases of repair is the ability to **find where the fault lies**. This will obviously improve with knowledge and experience.

You're going to have to quickly pick up the art of **positioning household appliances** (such as boilers, sinks, baths and showers) in the best possible places to ensure they function most effectively. Again, experience will help you greatly.

As you're also responsible for seeing that water can flow freely off the roofs of buildings, to avoid it becoming dangerously trapped, you must learn the skill of **shaping materials** to make this happen. This will often mean cutting lead to fix the roofing and carrying out repairs to the guttering.

It's not enough just to be able to perform the basic skills. As you'll be operating on your own in most cases, you have to be able to **plan your work** accordingly, devising a scheduled order of tasks to make sure you complete the job efficiently and effectively.

Different systems have operational quirks that you should learn and master before you take on their maintenance and repair. Since new systems are coming on to the market all the time, you will need to continually **re-educate yourself** to keep your product knowledge up to date. 'At present there isn't actually a qualification in renewables out there,' says Danny Davis, 'but training in renewables is going to be beneficial to any plumber because people who have knowledge of these new systems are going to be at an advantage.' His advice is to continually take courses in these new systems as part of your ongoing professional development.

RENEWABLES

This is energy from sources that won't run out such as sunshine, water and wind. Renewables are becoming increasingly important because they are environmentally friendly. The government offers a system of grants for householders who install renewable energy systems in their homes. For instance the grant for installing thermal solar panels is £400, about 10 per cent of the installation cost.

For more information visit
www.lowcarbonbuildings.org.uk

It will help you greatly if you have the **ability to draw** well – you'll certainly be using this skill a lot in the job, so it's bound to improve. You'll also need to be adept at interpreting the drawings made by others, checking that they are functionally sound and then putting them into effect.

All the work you do will require you to be able to **take quick and accurate measurements** consistently. Carry pencils and paper at all times to mark down the measurements you make.

You'll need to know how to use a **full range of tools and equipment** such as hacksaws and blowlamps, power drills and wrenches, not to mention different kinds of testing and measuring equipment.

As with the interpretation of drawings, you'll find it helps enormously if you're able to **interpret instructions** from others quickly and clearly. Sometimes you'll think that being a mind reader would be pretty useful in the job!

You'll need to be able to **work in all locations** and in all climatic environments. Although the bulk of the work is inside buildings you'll also have to brave the elements and work outdoors sometimes.

The primary purpose of plumbers is to maintain a flow of hot and cold water for each building, so your aim is to keep pipes clear for the swift removal of all waste products and water from toilets, baths, basins and sinks. Efficient drainage depends on you being able to **maintain the underground flow** of water and waste products into the sewerage systems.

Many plumbers also deal with heating systems that are fuelled by gas, oil and solid fuel and other systems such as cookers driven by the same utilities. If you work with gas you must, by law, be **registered with the Council for Registered Gas Installers (CORGI).** We talk about this in more detail in the Resources section of this book.

You'll need to be prepared to work not only on the guttering to ensure rainwater runs away smoothly but also on the roofs themselves to **maintain fully effective weather proofing** using such materials as sheet lead. Here, you'll have to make sure that overflow systems and cold water supply from the mains both continue to function properly.

You'll find that some plumbing companies specialise in domestic house work (Chapter 1 is an example of this), or alternatively concentrate on industrial and commercial work in shops and factories, while others may offer a broader range of services. Plumbers can work in quite a few houses each day on different call-out jobs or remain on a larger site for several months at a time. Only on rare occasions will you work at a fixed location (if you're employed to maintain systems at a hospital site, for instance).

Finally, the conditions you work in can vary massively. You could be working in a cramped and boiling hot loft in summertime, or on a freezing cold and exposed roof fixing guttering or frozen pipes in winter.

By now you should have a better idea of just what it is a plumber does. If you think you can handle all the situations and different aspects of the work described here then plumbing may well be the kind of flexible job for you. In the next chapter we will explore in more detail if you have the particular skills it takes to be a success in plumbing.

Tools of the trade

As a trained plumber you will find yourself using all kinds of specialist tools during the working day, but the most important tools you have are not those in your toolbox, but those you carry around within you. Some people are just not suited to a lifestyle where they are doing different things in different locations and prefer a more structured environment. Still others love the thrill of rising to a challenge, so a job where problem solving is involved is just what the doctor ordered. Listed below are just some of the skills – or tools – you will need to rely on as a plumber. Run through the list and see if you can recognise whether you have these particular strengths and skills. If you don't, they can be learned; but if you already possess them they will make becoming a plumber a much easier task for you.

- It may seem obvious to say so, but it really is true that to succeed as a plumber you have to be **good at working with your hands**. You will be sawing, wrenching, screwing up and unscrewing connecting pipes and joints every day in this job. There are all sorts of exercises you can do to improve your hand–eye coordination, but usually there has to be a certain amount of talent already there for you to work on.
- If you are good at DIY work around the house, or have an ability to carry out handyman and electrical repair work, plumbing may well suit you. If you feel that you could

improve the work you do with your hands, try **refining your skills** by doing some odd jobs in preparation.

- Being **conscious of safety aspects** is an absolute must for anyone working in the plumbing industry. 'All our students should have an awareness of health and safety issues,' says Tony Beaumont. Remember, you will be ensuring the efficient delivery of clean drinking water to people's homes or, if installing a central heating system, will be responsible for the gas and electricity connections, making sure they are correct and secure. Your own safety is also of utmost importance so following health and safety guidelines makes practical common sense.

- When you're learning your trade you will be working under supervision, and often that will mean carrying out jobs as part of a team. However this situation will not last forever – you will soon have to **rely on yourself**. A lot of the work that you will do as a plumber will be on your own so it is absolutely essential that you are able to inspire yourself to get jobs done. It can sometimes be very difficult to push yourself to do a job that is tricky, or smelly, or involves cramped conditions, but you must grit your teeth and do it. There will be nobody else around to do it for you.

- Plumbers will tell you that many of the jobs they do demand **incredible agility**. By definition, many of the water pipes, tanks and boilers will be hidden away and relatively inaccessible and often you will need the skills and flexibility of a contortionist to get at them! Natural fitness is therefore important. So, too, are **strength and stamina**. New baths, sinks, heaters and piping can be incredibly heavy and cumbersome to lift, so make sure you're up to the job by looking after yourself physically.

- In addition to physical stamina, it's important that you have the **mental stamina** to be able to cope with working in a pressured environment. The pressure of dealing with exams may be the only direct experience those of you new to a working life will have. Most of you will be able to manage easily and some may even prefer working under the added stimulus of pressure.

- Being able to **communicate well** with people is vital. First, you have to have good listening skills to really understand what it is that customers are telling you, because often their descriptions will be either very basic or confusing. You will have to interpret what they're saying and get to the root cause of the problem. You must also be able to explain very clearly to customers what exactly it is that you're going to do (or have done) and how much it's going to cost. This is extremely important because if a customer is pleased with the job you have done they will be inclined to use you again and to tell all their acquaintances about you. Repeat business and good word of mouth mean a plumber is guaranteed a large and loyal customer base, which is why looking after your clients well is so important.

- Even if you're working on your own you'll still be in **frequent contact with others**. You'll have to demonstrate that you can get on with people at all levels because you're going to be dealing with a constant flow of colleagues, customers and enquiries from potential customers. So, in addition to the fact that you need to be able to communicate clearly, you will also need to be sensitive to the needs of others.

- Because plumbing's a trade where communication has such a clear priority, you'll discover that having a **good sense of humour** is a great advantage. Sometimes, when things have gone wrong (as they invariably do in any walk of life), it's an absolute necessity. Plumbers are well known for their black sense of humour when having to deal with burst pipes, overflowing toilets and broken boilers.

- Being **reliable** and making sure you're always in a position to do what you say you're going to do is an absolute must. Closely linked to this is turning up when you say you will. These days people have busy lives and many will have to take time off work in order to let you into their homes. No-one wants to waste a whole morning waiting for a plumber who never arrives. If you are running late, ring your client and give them a new estimated time of arrival. Obviously when you're working for other people you also have an obligation to them to carry out all the work on time and efficiently. If problems do arise (e.g. parts not being available or a job taking longer than you quoted for) you must talk to your client immediately so that you can agree on a course of action.

- Strongly linked to reliability are the personal qualities of **trustworthiness and honesty**. You will often be working alone in customers' homes and they will need to feel sure that they can leave you to work unsupervised. You must, therefore, be able to inspire their trust. Providing references and becoming a member of an association such as the IPHE are both good ways of reassuring clients. Also, don't just have a mobile phone number, but give people an address and landline they can contact you on as well. Again, you'll find it pays dividends in increased business as word spreads on the grapevine.

- Just about every job you'll do as a plumber requires you to **travel quickly** from A to B (unless of course you're working on installing systems on some large new project), so you will have to learn to drive. This is especially true if you work on your own and have to transport all your tools, hardware and equipment with you.

- A lot of the work involved in plumbing is painstaking and delicate, demanding a **great deal of patience**. You may be engaged in a particularly tricky aspect of fitting a system, or trying very hard to replace an especially small and inaccessible part. It's not good enough to be almost right so take your time and make sure it's perfect first time.

- You will have to learn to **maintain your cool and be polite** at all times. However unreasonable your client may be, keep your temper and explain everything clearly and logically – to colleagues as well as to customers. **Cleanliness** goes hand in hand with politeness. Customers do equate a good appearance with reliability and honesty. Even if you know you are going to get filthy on the job, arrive looking neat and tidy and keep a dust sheet and a change of clothes in your car or van.

- If there is one subject that all plumbers are likely to use a lot, it is **mathematics**. Whether you're working out angles, totting up sales figures or calculating your wages, good attention to your maths studies at school and college is bound to help in this career. The same is true of **scientific knowledge**. Anyone who aspires to a job in plumbing should have a sound knowledge of the basic principles of chemistry and physics – and if you want to get involved in design and engineering you'll have to take your understanding of theory and practice to a much higher level. You must also have **good reading skills** to read and understand technical instructions.

- Finally, you must have the ability to **adapt** to different and changing circumstances. No two jobs are exactly the same and you have to be willing to learn about new aspects of your work all the time. You may also have to be **flexible** about your hours of work. If you're on a 24-hour emergency call out service, then you have to be ready to carry out a job at any time, which can play havoc with your social life.

Possible problem areas

Just as there are certain skills or personal tools that will help you get on in this industry, there are also certain physical conditions that could hold you back, and certain aspects of the job that may not suit you. Think about these seriously if you are considering a job in plumbing.

- If you suffer from **breathing difficulties** or have a condition such as **asthma**, working in plumbing may not be for you. The dust, damp and changes in air temperature that a plumber comes across can exacerbate existing breathing problems. Carry any alleviating sprays or medication with you wherever you go.
- Similarly, if you suffer from **claustrophobia**, you're going to find some of the work very tough indeed. As mentioned already, many boilers, pipes and water tanks are situated in confined spaces with severely limited access – if you don't like the idea of lying under the sink with your nose in the U-bend and a crick in the neck, then you're probably going to be in trouble.
- If you have a family history of **back problems** the constant lifting and carrying of heavy equipment and hardware could be storing up trouble for you in the future. This is a physically strenuous job and even though you will be taught the correct and safe way to lift and carry at college it is best to bear in mind that many people are forced to

leave the industry owing to bad backs. As Tony Beaumont says: 'As you get older you don't want to keep carrying radiators up flights of stairs.' In Case Study 3 with Paul Tudor (see pages 51–3) he explains how the profession is also very hard on your knees, and arthritis in both knee and hand joints is an occupational hazard for plumbers.

- You may have difficulty if you suffer from **colour blindness**. It is advisable to check with organisations such as SummitSkills (see the Resources section for more information on how it may affect your chances of becoming a plumber.

- Some parts of a plumber's job require you to work outside, maybe on roofs to deal with pipes and guttering, and if you suffer from **vertigo** you should be aware that this might occasionally cause you some distress. If the idea of huddling on the roof of a three-storey building in snow installing a new central heating ventilation unit gives you a panic attack, then you may have to think again about plumbing as a career.

- If you're **no good at organising yourself**, then you're going to have to learn fast. Plumbers are always on the go, usually dashing from one job to the next, so you'll need to have various essential support materials ready and in good working order.

- The flipside of punctuality is **poor timekeeping**, and if you're someone who always seems to be late then it's a problem you're going to have to solve very quickly. This shouldn't be too arduous – you'll just have to learn to aim to arrive ten minutes early in the morning and to try to keep that up throughout the day.

By now you should have a pretty good idea about what a job in this industry actually entails and what it takes to be successful in plumbing. If you still believe this could be the career for you, then complete the short quiz below. The answers will enable you to see if you are really suited to working in this industry. Just tick either yes or no to each of the questions.

QUIZ

Do you like to see the results of your work immediately?
YES/NO

Do you believe that variety is the spice of life?
YES/NO

Do you think customer service is really important?
YES/NO

Do you have the patience to take great trouble over little things?
YES/NO

Do you think teamwork is a vital part of the job?
YES/NO

Do you find the prospect of working in an office a bit depressing?
YES/NO

Do you enjoy working with your hands whilst still using your brain?
YES/NO

If you answered 'YES' to all or most of these questions, then you've probably got the right instincts and temperament to be a plumber.

If you answered 'NO' to most of these questions, then you will probably find a career in a different industry is more suited to you. It might be a good idea to check exactly what you want to do and look at other career options.

If your answers are fairly evenly split between 'YES' and 'NO' you may want to think very carefully about everything that being a plumber means. Read on, but be ready to make some changes to your expectations.

A day in the life of a plumber

A busy plumber talks us through his working day and answers other questions about his job.

Steve Evans is a mature trainee, who switched careers reasonably late in life in order to take up an apprenticeship as a plumber. He works mainly in the London area.

HOW DID YOU GET YOUR FIRST JOB – AND WHAT WAS/IS THE TRAINING LIKE?

I got my first job as a follow-on from my first training placement. The placement lasted ten weeks, which gave my prospective employer plenty of time to find out whether I was punctual, diligent and could work at a reasonable speed, and he still offered me a job working with him!

I didn't get paid very much (about £200 per week) but I was lucky to be getting broad experience rather than simply doing central heating, for example. It was difficult for me, aged 36, to find anyone who would train me. I did a lot of research, most of which led to disappointment. However, through LearnDirect I found out about a National Vocational Qualification (NVQ) course at the Community Refurbishment

Scheme in Deptford, London. Amazingly, they not only interviewed me but also offered me a place. The tutors were good and very knowledgeable. They teach you how to bend pipes accurately, prepare pipe ends perfectly, ready for soldering, work out the precise fall (downward slope) required for a certain length of basin waste pipe and so on.

Then, when you go on site and are spotted measuring fastidiously or 'wire wooling' a pipe end, you get shouted at for taking too long. Doing the NVQ, I discovered, was very much down to my own commitment to get on and complete the necessary practical and theory work. Several people on the course simply did not do this and ended up not being able to sit the exam. That was sad because nobody was incapable of doing the work.

WHAT HAS HELPED YOU TO PROGRESS?

It sounds stupid, but working with someone who realises you want to do plumbing helps you to progress really rapidly. My boss threw me straight in at the deep end, plumbing in bathrooms and kitchen sinks. Some trainees spend at least a year carting tools and fittings around and generally labouring before they get a look-in on a blowtorch. On

On placement, make sure early on that everyone knows you're a plumber

placement, make sure early on that everyone knows you're a plumber, not a labourer. That doesn't mean you say no to jobs that don't involve pure plumbing, it just means you want the opportunity to try the real plumbing work.

CAN YOU SUMMARISE A TYPICAL WORKING DAY?

When you get put on a placement you realise that the job is 95 per cent 'heavy grind' labouring: shifting boilers, radiators, baths, toilets, etc. Removing the old stuff, ready for replacement, is often the most work, especially if it involves taking out a hot water cylinder.

A typical working day (starting at 8am) for me, say on a refurbishment site, might be fitting a bathroom followed by a kitchen sink. This usually involves around 20 trips to the plumbing store (often a big metal container) and skip, which are rarely conveniently sited.

In the morning I start by 'ripping out' the old bathroom: knocking tiles off the walls; isolating the water supply; cutting pipework; and removing the old heavy bath, basin and toilet (yes, they are mostly disgusting to handle) and placing them in the skip. This involves a fair amount of water and mess, which then has to be cleaned up. I fetch the new heavy bath and lug it up two flights of stairs, tripping over various obstacles and squeezing through narrow doorways. After this, I 'make it up' with feet, taps, a waste and overflow and, often, bath handles. I then set the bath into position on 2" × 3" timbers, and level it.

After that, I plumbed the bath with copper pipe, soldered fittings and welded waste pipes and fittings. Most of this time I am lying on my back on a dirty floor, or kneeling, or bent double. With that finished, I fetch the basin and make that up. I place it on its pedestal and get it to look level (unlikely that a spirit level will help you with this because basin moulds don't seem to be very accurate). With it then screwed to the wall I plumb it in. This is followed by the toilet, which rapidly reveals to me that the floor is very far from level and a plywood plinth must be constructed. In the afternoon, having waited for a kitchen fitter to fit the sink base unit, I do the same to the kitchen sink and hook up the washing machine and dishwasher too. Going-home time is usually around 3.30pm. (*It is important to note that this is Steve's personal experience; the length of a working day will vary greatly according to where you work.*)

WHAT ARE THE BEST (AND THE WORST) THINGS ABOUT YOUR JOB?

The worst things about the job are the dirt and grubbing around on the floor or even underneath it; the pain of getting soldering flux into one of the many open wounds on your hands; and the aching back and knees. The best things are getting ahead of yourself on a price-work contract and being able to go home regularly at 1.30pm; finishing some pipe work and knowing that (a) it looks tidy and (b) it doesn't leak; not having to take work home with you; and feeling pleasantly physically tired, not mentally distressed.

WAS IT EASY TO SWITCH CAREERS, AND HOW BEST SHOULD OTHERS GO ABOUT IT?

Switching careers to plumbing depends on your age and what you have been doing up to that time. It was not at all easy for me to find an opening on a course, and another way to do this at my age would be to save up some capital (to live on) and beg a plumber or plumbing company to let you work for free for two months. If they like you, they'll work something out.

Think very carefully about whether you really want to
do it. If you do, you should find there are plenty of
openings through the local job centre

HAVE YOU ANY ADVICE FOR SCHOOL LEAVERS WISHING TO BECOME PLUMBERS?

My advice for school leavers is this: you should think very
carefully about whether you really want to do it. If you do,
you should find there are plenty of openings through the
local job centre. Then put a lot of commitment into it and
it'll be plain sailing. It's not easy work or easy money,
despite what people think at the moment. However, it
can pay quite well and it will offer you a great deal of
satisfaction.

CAN YOU IDENTIFY ANY QUALITIES THAT MAKE PEOPLE MORE LIKELY TO SUCCEED?

If you don't mind getting dirty and are not over-fussy about
other people's dirt, and if you're physically co-ordinated and
mentally capable of rudimentary maths, then you'll get on. If
you're relatively strong or persuasive enough to get others to
help you and if you don't mind the grind of labouring which
is, as I said before, most of the work, then you'll succeed.
Also, you shouldn't be afraid of trying something new or, at
least, something you haven't done before. Even a qualified
plumber will wander into unfamiliar territory. If you're
sensible, able to read instructions (e.g. for an all-singing, all-
dancing jacuzzi bath), and confident of your general abilities
to adapt, then you'll go far.

WHAT DO YOU THINK THE FUTURE HOLDS FOR YOURSELF AND THE INDUSTRY?

Well, there'll always be work, that's for sure. Water damage frightens people and most will always want to be rescued by a plumber. It's not complicated, but there are some key considerations, second nature to a plumber, which may elude the DIY-er entirely. Personally, I'd like to see plumbing brought into the 21st century with the spread of solar panels for hot water and heating. This is possible now with new hyper-efficient panels, but people, and often especially plumbers, don't know it.

Making your mind up

6

It's almost crunch time – do you feel you've gathered enough information about plumbing to commit yourself to training for the qualifications that will help you be successful? Just before we examine the ways in which you can get into the industry, here are a few of the most commonly asked questions about training for a job in plumbing and what you, personally, can expect to get out of a job in this industry.

WHAT ACADEMIC QUALIFICATIONS DO I NEED TO GET ONTO A TRAINING COURSE?

Even though there are no specific academic requirements for you to be a plumber, you do obviously have to have a high level of mental alertness and, as places on training schemes are currently at a premium, it's best if you can achieve reasonable exam results (for these purposes Grade A–C GCSE for entry into an Advanced Apprenticeship) in English, Maths, Science, Craft and Design and Technology.

'People think of vocational trades as being for students who are not academic,' says Tony Beaumont, 'but that's really not the case. It makes it so much simpler if students have Grade Cs in English, Maths and Science, both for them and for us as trainers.'

Danny Davis agrees: 'We need people who not only have good Science, English and Maths qualifications but also those people who can use their brains as well as their hands. It's always good to have an engineering mind, to understand how things work and to have a thirst for knowledge.'

Some employers may ask you to take a selection test to see if you are a suitable candidate for working in the plumbing industry. This is to help you as well as your potential employer because there is no point in putting yourself through the training if you are not cut out for all the aspects of the job.

HOW LONG DO I HAVE TO TRAIN FOR?

There is no set length of training because, on the one hand, you can choose to take the qualifications at whatever speed you wish, and, on the other, it depends on how far up the levels of qualification you wish to go. However, if you're considering trying to pass the NVQ Level 1 or 2, it should take two to three years, and progressing to Level 3 should take a further year to 18 months. Training for such qualifications will be discussed in detail in Chapter 9.

WHAT'S THE PAY LIKE?

Good. 'Being self-employed you can earn the lion's share of the money,' says Tony Beaumont. 'However, you will have more job security working for a large company.'

As a general guideline, you can expect to be earning an annual salary of £25,000 after you have been employed by a company for a period of two years. Whilst you're qualifying you would obviously be earning less than that. First-year apprentices may earn as little as £5,000, rising to £17,000 a year when newly qualified. If you're working unsociable hours (such as on 24-hour call out emergency services) you can expect to be paid at a higher rate to compensate for the disruption to your life.

Those of you who progress to becoming self-employed can expect to earn between £30,000 and £40,000. There is a great deal of talk about plumbers earning salaries of around £100,000, but this is highly unusual and depends on other factors such as working anti-social hours and running your own business. The plumbing industry also has a pay-related pension scheme, which means that it offers a level of job security that cannot be found in all industries.

DO I HAVE A NORMAL WORKING WEEK?
Yes, you usually do, varying from eight to ten hours each day at the normal times, depending on the terms of your contract.

However, if you choose to work on emergency 24-hour call out (or if that's the type of work that you get offered) then you will naturally have to be more flexible.

WHAT ABOUT HOLIDAYS?
Holidays are standard, offering you on average four working weeks as holiday each year, as well as all the public holidays available (Christmas, New Year, Easter, Bank Holidays, etc). Many companies will also offer extra days as holiday depending on your length of service, so your entitlement may rise to five weeks, or above, the longer you stay with the same employer. If you work for yourself, one of the great advantages is that you can choose your own holiday arrangements, but remember, if you are self-employed,

no-one pays for your holidays and while you are taking time out relaxing you will not be earning. You should bear this in mind when working out how much you need to earn a year.

CAN I CHANGE CAREERS EASILY?

Yes, you can. As you will have discovered, many of the skills that you learn in the industry are readily transferable, and your knowledge of water might give you a great start if you wanted to build a business installing and maintaining swimming pools and hot tubs. You've also learned flexibility, such as the ability to work on different systems that might enable you to succeed as an electrician. The work that you've done on roofs could translate into finding a job in that part of the construction business – indeed, a great many of your skills would fit well for a job in that industry.

Promotion prospects are very good indeed within all areas of the industry

CAN I WORK ABROAD?

You can work in any country that is a member of the European Union (although obviously being able to speak the relevant foreign language will help). If you want to try this out there is a Young Workers' Exchange Programme (for 18–28-year-olds) that will give you work experience or training in the country of your choice, for as little as three weeks or up to 16 months.

There are no such guarantees for other parts of the world, but if you have some solid qualifications behind you they should stand you in good stead.

WHAT ARE THE PROMOTION PROSPECTS?

Promotion prospects are very good indeed within all areas of the industry. Obviously a great deal, if not all, depends on you. If you work and study hard, complete your training swiftly, develop your expertise and show willingness to do more than is necessary, you'll get on just fine. You'll find a discussion of some typical promotional career paths in the section on Career Opportunities (see Chapter 11).

WILL I GET HELP IF I WANT TO BECOME MY OWN BOSS?

Yes, there are several booklets and courses available to help you prepare for the various facets of setting up your own company. We will discuss these in the Career Opportunities section (see Chapter 11).

WHAT DOES THE PUBLIC THINK OF THE INDUSTRY?

Frankly, the popular perception of plumbing is erratic. Anyone who has had a bad experience or been ripped off by cowboy plumbers will naturally tend to think negatively, whereas those who've had swift, efficient and friendly service will focus on the positive sides of the profession. According to a recent consumer survey carried out by the Institute of Plumbing and Heating Engineering (IPHE), 59 per cent of those taking part thought that plumbers provided an essential service, and only 6 per cent said they were undervalued, with a further 6 per cent saying they thought all plumbers were cowboys.

DID YOU KNOW?

The actor who plays Phil Mitchell in 'EastEnders', Steve McFadden, used to be an apprentice plumber.

There is also a certain amount of misunderstanding about how much plumbers actually earn. Some people seem to think that all plumbers are raking it in (in the same survey 29 per cent of people thought plumbers overcharged for their services), when the sober truth is that those who are earning most are those who are working all hours of the day and night. Attitudes are changing slowly and of course it will partly depend on you (and others like you) to restore and maintain plumbing's good name.

WHAT WILL I GET OUT OF IT APART FROM A CAREER?

Basically, you'll get three main things – a strong feeling of satisfaction and self-worth, a chance to meet a broad cross-section of people, and the peace of mind from knowing that your own household plumbing dilemmas are solved for life.

You'll feel satisfied because you are able to help people by solving a constant flood of problems, each of which is of vital importance for them. To be able to put things right will give you a powerful sense of achievement. Furthermore, you'll invariably be meeting many different kinds of people. As in the rest of life, you probably won't want to be friends with them all, but some you may, and communicating with each and every one of them will be interesting on many different levels.

7

ELLEN CHEESEMAN

Case study 1

Ellen Cheeseman is one of a small but growing number of female plumbers. She works mainly in the south-east of England.

How did you get your first job – and what was the training like?
I worked with a qualified plumber whilst training then started doing small jobs for myself. It was really difficult at first, knowing what to charge and remembering everything, but once I'd made a big mistake and had to call the same friend in to help me out, everything was fine after that!

What has helped you to progress?
Having the determination to succeed. This may sound corny but it's true, because I don't like to be beaten. I also got involved with my local branch of the IPHE, which I found to be very beneficial.

Can you summarise a typical working day?
No, every day is different. That's what I like most about being a plumber!

What are the best (and the worst) things about your job?
I think the best part is working for myself and seeing the look on satisfied

Every day is different. That's what I like most about being a plumber!

customers' faces once a job is complete. I can also work the hours I choose so that I am around for my son. The worst bits are all the hours you work for no pay, behind the scenes. It takes ages to keep up with all the paperwork.

Was it easy to switch careers?
No, not particularly, but if you really want to and can get someone to give you a push, it's possible. I had a lot of support from my family, which helped. Also I wasn't in the position of having to work to contribute to the household.

Have you any advice for school leavers wishing to become plumbers?
Aim as high as you can. Work hard to get all your qualifications and behave like a professional right from the beginning. If possible, try to be multi-skilled by taking on an extra course in plastering or decorating. This will give you many advantages in the future, since your skills will always be in demand. Go for it, it's worth it in the end.

Aim as high as you can. Work hard to get all your qualifications and behave like a professional right from the beginning

Can you identify any qualities that make people more likely to succeed?
The two most useful are determination and courage. Also you must have a belief in what you're doing and in yourself.

What do you think the future holds for yourself/the industry?
Who knows – I'd like to build up a large company providing

all the services possible to do with plumbing, employing an equal number of women and men plumbers. There are also opportunities to go into teaching plumbing courses. At the end of the day, I believe there will always be work for a professional plumber.

8

JENNY ELTRINGHAM

Case study 2

Thirty-year-old Jenny wasn't always interested in plumbing, but she has always been practical and didn't like working in an office environment. She studied for GCSEs and A-levels, and then did various office jobs before deciding to switch to plumbing.

She started out by working with a friend who was a plumber but then decided to go to college to qualify. She did the City and Guilds technical certificate and NVQ Level 2, 3 days a week in college and 2 days a week working with the local council. She then worked with a building and maintenance company before joining Pimlico Plumbers in London as a plumber. She is now working towards her CORGI gas qualifications, which will enable her to install and maintain boilers, therefore broadening the potential of her job considerably.

'On a day-to-day basis I will usually undertake a variety of jobs, supplied through the company I work for. I work generally between 8am and 6pm, Monday to Friday. My jobs, as a general plumber,

I like being free from office life, being in charge of m own jobs can be anything from repairing leaking pipework to installing new sinks, washing machines, showers and taps. I like being free from office life, being in charge of my own jobs, the satisfaction when you crack a particularly awkward job, and the problem-solving aspects of the job. No job is the same and each job seems to have some hurdle to have to think around (e.g. having to get access to a leak, or how to fit something into a bathroom with seemingly no straight walls or floors!)'

Jenny says the only downsides to her job are the occasional awkward customers and it is sometimes difficult to predict how long a job will take, so working hours can be long as most of the time it's not possible to leave a maintenance job unfinished. Because of this her social life has suffered a bit.

Ultimately, Jenny would like to swap the stress of working in London to set up on her own outside the city. She says the qualities you most need to succeed as a plumber are patience, calm and practical problem-solving abilities, and good communication skills.

'I'd say to anyone wanting to do this as a living, do a proper apprenticeship, working with an employer in conjunction with college. Practical experience is irreplaceable. Do not do a crash course – it is impossible to learn plumbing in 8 weeks!'

Training day

9

By now you should have a good idea whether or not plumbing is for you, but before you can go out there and start to work you will have to undergo training. As stated previously, changes to building regulations that came in last year (2006) will mean that within the next two years all plumbers will be required to be 'competent' and so will have to attend and pass a Competent Plumber Scheme (CPS). As the CPS is still being formulated there are no details as yet, although it is highly probable that knowledge of renewables (e.g. solar energy and ground source heat pumps), and energy and water efficient technologies will be on the agenda. In the meantime – and even when the CPS is introduced – the information given below on NVQs and Apprenticeships will still be relevant.

Although there is no upper age limit for training, and there are now many older people who have decided to give up their initial career and retrain as plumbers (apparently the average age of the IPHE's 11,000 trainee members is 35), in order to get funding for most courses you must be under 25 years old. That is why going straight from school into a course or an apprenticeship makes good financial sense. (If you are older you may be eligible for a Career Development Loan – see page 48.)

NATIONAL VOCATIONAL QUALIFICATIONS

The main plumbing qualifications are the National Vocational Qualifications (NVQs, or Scottish Vocational Qualifications –

SVQs – in Scotland). Most colleges will prefer you to have Maths and Science GCSE grades A–C to take NVQs. NVQ/SVQ Level 2 gives you a basic foundation in plumbing, but the IPHE recommends all plumbers to attain Level 3, which is more comprehensive and deals with domestic, commercial and industrial plumbing. The IPHE is also runs a Master Plumber Certificate, for those who have already attained Level 3. NVQs/SVQs allow you to learn practical skills on the job whilst training at college on a block or day release basis.

In all there are five levels of NVQ, with Level 1 being the most basic and Level 5 the most advanced. This means that the skill levels in NVQ/SVQ terms can be broadly defined as:

NVQ/SVQ Level 1 – Basic skills, with an introduction to your chosen topic

NVQ/SVQ Level 2 – Completion of Foundation skills

NVQ/SVQ Level 3 – Advanced skills, to give you specialist knowledge

NVQ/SVQ Level 4 – Supervisory skills, or very advanced technical skills

NVQ/SVQ Level 5 – Managerial skills for those who want to progress further.

APPRENTICESHIPS

If you are aged between 16 and 24 you are eligible to do an Apprenticeship (Skillseekers in Scotland). Apprenticeships and Advanced Apprenticeships allow you to earn as you learn. You must have at least four GCSEs at grade C to apply. The Advanced Apprenticeship leads to the equivalent of NVQ Level 3. In both cases you have to do some on-the-job training with a company and some theoretical training at a local college.

You can do this at the same time or in stages. So you'll take one of three routes:

1. ATTEND A COLLEGE AS A PART-TIME STUDENT
You usually go either one day a week or in weekly blocks. You can try to pass the theoretical side of the qualification as fast as possible to help you move on to the practical stage of working with a company.

2. JOIN A TRAINING PROGRAMME
If you can, it's best to do the practical and theoretical stages at the same time because it helps you to relate all the pieces of learning to each other. Training programmes will be run either by an employer or by an organisation that works with employers to provide training such as SummitSkills (see Resources chapter).

You will be helped to build up a National or Scottish Vocational Qualification in all the necessary subjects at the right NVQ/SVQ levels for your chosen career (usually a combination of Levels 1, 2 and perhaps 3).

3. GO STRAIGHT INTO EMPLOYMENT
If you cannot find a position on a training course that suits you, it might be possible to join a company, establish yourself in the job, and then start your formal training later.

Once you have qualified as a plumber you can continue training to reach NVQ/SVQ Levels 4 and 5 if you wish. You can also take qualifications in related subjects such as welding and electrical installation (and all the other skills related to Building Services Engineering) that have appropriate NVQ/SVQ levels.

WHAT TRAINING COURSES EXIST?

Training courses, as stated earlier, are a mixture of on-the-job training from the company you are working for and off-the-job, theoretical training from a local college. There are many of these colleges in all parts of the UK.

You can find lists of possible companies and colleges on the IPHE website. There are also full details of colleges and training centres on the SummitSkills website and the British Plumbing Employers' Council (BPEC) website. All these addresses and lots more useful contact information are contained within the Resources chapter of this book. Alternatively, scour your local phone book for possibilities. It's probably a good idea to keep a record of the progress of all your enquiries.

There are five main ways of being taken on for training:

- you are nominated by national training providers;
- you are nominated by local training providers;
- you are nominated by careers services/Connexions;
- you are nominated by your local employment agency;
- you are nominated by your employer.

All routes require you to take some action on your own behalf.

NATIONAL TRAINING PROVIDERS

Each country in the UK has a designated national training provider for the profession. SummitSkills is responsible for skills development throughout the UK and, in conjunction with various national bodies, forms these plans.

LOCAL TRAINING PROVIDERS
All the countries are subdivided into smaller areas, where the advisers will obviously know in greater detail what is happening in your locality. Look at the Learning and Skills Councils for England, Scottish Enterprise and their Local Enterprise Councils for Scotland and, in Wales, at Education and Learning Wales.

CAREERS SERVICES
Both your local and your school careers services can refer you for training in the profession if you can demonstrate that you have the potential to succeed.

CONNEXIONS
Connexions is the government's support service for all young people aged between 13 and 19 in England. It can give you advice and help on starting your career and offer you personal development opportunities.

LOCAL EMPLOYMENT AGENCY
Your Job Centre or local employment agency can refer you for training in the profession. Again, you must show you have the potential to succeed.

EMPLOYERS
You can take the initiative and, if you are already in work, convince your employer to apply for you to receive training. Alternatively, you can approach prospective employers and undertake to carry out the training as a condition of employment.

THE NEW DEAL

The New Deal applies particularly in an area where job opportunities are limited. The scheme supports you if you're having difficulty in finding work by helping you get training towards a particular career.

If you decide to participate in the scheme you will have the support of a New Deal adviser who will help you to find a job and get the appropriate training. This training will normally lead to a part- or full-time NVQ.

The New Deal scheme is administered through the Employment Service. To find out the contact details of the service nearest to you, it's best to refer to your telephone directory or your local careers office.

BUILDING SERVICES ENGINEERING NVQS

SummitSkills is offering two NVQs in Building Services Engineering Technology and Project Management. They are available at Levels 3 and 4, and are based on national occupational standards developed by SummitSkills and awarded by EAL, an awarding body for engineering vocational qualifications.

Both qualifications offer progression from the skilled trades into office- or site-based technical/supervisory/managerial posts or recognition of these competences within:

- air conditioning and refrigeration;
- electrotechnical;
- heating and ventilating;
- plumbing.

Typical candidate job roles include project/contract manager, supervisor, estimator, sales engineer, site/workshop manager and design engineer. (For more information contact SummitSkills – see Resources.)

MATURE TRAINEES

Because so many Mature Trainees (generally defined as anyone over 24) have been expressing interest in trying to switch career to become plumbers, there is now an accepted pathway for you to follow.

As a Mature Trainee, you can expect to achieve the same qualification as young people proceeding through an Apprenticeship framework, which leads to achievement at NVQ Level 3.

This would include passing the knowledge content of the Certificates in Plumbing, picking up experience in the workplace and ensuring that you undertake the appropriate upgrading of Key Skills in communication, application of number and information technology to match the requirements of a plumber.

You'll be pleased to hear common sense prevails here. If you are considerably older or have been involved in a related profession (say, construction work) then you'll find the time you need to finish training will be reduced. This will obviously depend on the match of the range of existing skills – the more you've learned already, the shorter your training will be.

However, most of you will not be able to find full-time college places. Entry for many of you will be achieved by proceeding straight to employed status. This would mean your working with an employer and attending a centre (normally a college) on a part-time basis.

Naturally you will be given targets to attain. You will have to pass the two stages of certification – the Certificate in Plumbing (Intermediate) and the Certificate in Plumbing (Advanced). Your employer will provide the work-based part of the programme and the training route will normally be managed by a Learning Provider, who will receive the training funding from the local Learning and Skills Council (or from a Regional Authority in Scotland and Wales).

OTHER METHODS OF LEARNING

Some employers may have their own in-house training programmes that conform to the standards set by the training authorities and give you the equivalent of the NVQ/SVQ qualification.

Distance learning packages are also available, which give you all the academic content you need to complete that side of the qualification, although you will need to buttress it with some practical training to gain your NVQ/SVQ.

BPEC/PHIA APTITUDE TEST

This special test has been developed by the Plumbing and Heating Industry Alliance (PHIA) in conjunction with BPEC and is available as part of a free CD-Rom from SummitSkills. It will give you an indication of whether you might have the skills and qualities to succeed as a plumber.

DID YOU KNOW?

The average pay for a plumber can rise depending on the project he or she is working on. For instance, workers on the massive Heathrow Terminal Five project are earning up to £50,000.

When you have completed your aptitude test on the CD-Rom, record your score and the consequent result, and include a copy of the information with your application to any potential employers, not forgetting to keep a master copy for your own records.

If you are not successful on the practice aptitude test, seek advice from your local college on the best ways to upgrade your skills. Remember you can always prepare and practise so that you can take the test again.

FUNDING

You'll have noticed some references to funding earlier and the good news is that there is funding available to support your training and assessment, provided either through the local Learning and Skills Council (or Regional Authority if you live in Wales, Scotland or Northern Ireland), or from Job Centres if you are unemployed.

Once you've applied, the availability and amount of funding, the timescale over which it is provided, and when it will be paid, will all be identified as a result of your initial assessment by the Learning Provider. As you'd expect, some regional variations tend to occur but, whatever happens, you or your employer will be expected to make a contribution to this process. In other words, the funding will not completely cover your costs.

The Department for Education and Skills (DfES) offers a Career Development Loan for those undertaking a two year vocational training course. It is available to employed, self-employed and unemployed people and covers full-time, part-time and distance learning courses. Basically, the DfES pays the interest on the loan for the length of the course and up to one month after it finishes. Details are given in the Resources chapter.

The diagram opposite sums up the various routes into a career in plumbing, from leaving school right to the top of the profession.

BEWARE ROGUE TRAINERS!
Over the last few years, as a shortage of plumbers was predicted, many companies and individuals set themselves up as plumbing trainers charging thousands of pounds to train people to be plumbers in a matter of weeks. These fast-track plumbing courses do not have **recognised certification** and are not long enough to train a person comprehensively. So, there are now people out there who think they are plumbers but they would not be eligible for the upcoming Competent Plumber Scheme because they do not have the skills, experience or proper certification. The IPHE gets many calls from people who have seen these courses advertised and wonder whether to spend the money on them. The IPHE's advice is *don't*. You need to join a course that leads to a proper NVQ certification. To find a proper certified course contact the IPHE or SummitSkills (see Resources).

Information courtesy of Danny Davis (IPHE)

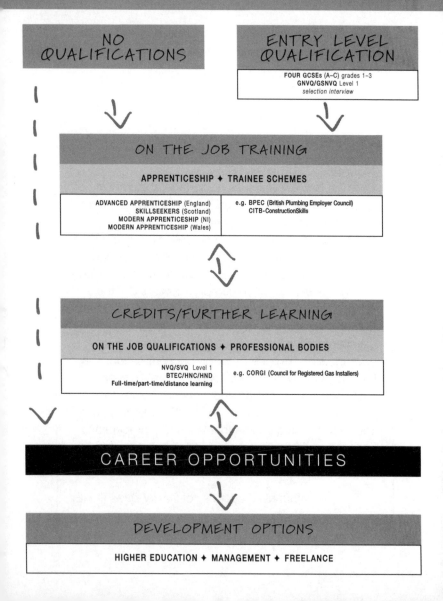

PAUL TUDOR

Case study 3

*Paul Tudor is a 37-year-old plumber and
ACS and CORGI accredited gas engineer.
He is currently self-employed and works in
the Kingston and Surrey area.*

'I started out as an apprentice with British
Gas back in 1982 and I worked for them for
seven years before switching to TRANSCO,
their safety division, and basically I worked
for them right up until four months ago
when I went self-employed.

'At school I didn't know what I wanted to
do so the careers adviser gave me some
brochures on plumbing and the fire
service. Well, the fire service wasn't
recruiting but British Gas was and so I
wrote off to them. At the time there were
85 people going for just two jobs and I
was lucky enough to get one of them.

'I was a typical kid who had never had any
interest in plumbing but this was a great
opportunity and I actually settled into the
role fairly comfortably. There were a lot of
other lads working with me who shared
the same sense of humour and so I did
feel at home. I started by doing domestic
appliances with British Gas and I liked the

The best
things about
my job are
the money
and the sense
of satisfaction
you get when
you do a job
well

fact you were never tied down to the same place any day because I just knew I didn't want an office job. I worked fairly regular hours until I transferred to TRANSCO and of course they have 24-hour call out because of emergencies. As an apprentice you work with another engineer, then you go out on basic duties for a probation period before being sent out on your own. Generally, when you first start with British Gas you are servicing boilers and when you start with TRANSCO you are resetting meters. As you progress you go on to boiler breakdowns and emergency stuff.

'I think the training has made me so aware of the dangers of gas. I was never nervous about working with it. It means you never go into situations where you don't know what you are doing. I suppose if you sat back and thought about it you might get nervous. I have had some hairy moments though. I remember one instance when I was working for TRANSCO when I was called out at 11pm to a block of flats. The smell of gas hit me as I walked into the block and when I walked into the lift I immediately got explosive readings on my detector where the gas had settled in the lift shaft. It turned out someone had left their gas cooker on full blast and everyone in the block could smell it except for them! British Gas and TRANSCO have to keep up with the latest health and safety standards and initiatives and that meant throughout my career with them I was sent on lots of refresher courses and training courses to keep in touch with all the new technology.

'I became self-employed mainly for family concerns and now I am working on a big project on a block of converted flats and I am subcontracted back to British Gas for three days a week. At the flats I am installing all the boilers and doing all the plumbing including showers, baths, sinks, toilets and absolutely all the pipe work. It's a massive job but I really like it because there is a big group of us working together and the banter is great. Also, the electrician is my brother-in-law and the plasterer is my father-in-law so we really are one big family.

'The best things about my job are the money and the sense of satisfaction you get when you do a job well. Say someone has gone out and bought a nice decorative gas fire and I put it in and it's a neat job with no pipes showing and I turn it on and leave them with it glowing up nicely. That's a lovely feeling to leave someone happy and cosy, that's really fulfilling.

'The worst thing is most of your life is spent on your knees. As you get older you really start to feel it and it really does take its toll because if you are doing pipe work you will generally be on your knees. Also, it's fiddly, intricate work and you really have to concentrate so it is mentally hard. Just think about the bottom of a boiler – you are going to have to connect a gas pipe, a vent pipe, a hot water pipe, a cold water pipe, and pipes to all the radiators and they all go in different directions so there is a lot of thinking involved. Also, people dying because of something I have or have not done is my greatest fear and so I try to do the job to a level of safety above and beyond the call of duty.

'Now I just want to make a decent living at what I do but if the opportunity of a big contract came up then I would start employing people. I'm finding I'm getting a lot of work through word of mouth and I think I get recommended because I always try to be neat and tidy, I'm safe and I'm legal and I always give people a card with my address on it so they know where to find me if there are any problems.'

Career
opportunities

11

Of course it's entirely possible that you may be more than happy to stay as a basic 'wet only' (i.e. no gas work) plumber for the whole of your working life. Some people are also quite content to stay with the same employer for most or all of their working days. However the possibilities and opportunities to advance up the plumbing career ladder are enormous and some of you may wish to move on and develop your career into broader fields. How far you take it will depend on how much time and effort you are willing to put into more advanced training and education. For instance, Tony Beaumont says: 'More academic 15-year-olds who don't want to get their hands dirty can branch out and go into the more technical design side of plumbing.'

With more experience you could **start your own business.** Because it is relatively simple to do, many plumbers who work in the domestic sector start up their own business, giving them the chance to earn greater rewards (but with greater risks and the likelihood of far longer hours). The lure of being your own boss is always attractive. If you prove successful, one day you could be taking on trainees or apprentices of your own. However, you must be prepared for all the extra paperwork that managing your own company will undoubtedly bring with it, not to mention the special problems involved in employing other people if you do

This is an industry with real opportunities for advancement. As you go through the training and discover where your strengths lie you will be able to map out a future career path. The diagram below shows options that will open up to you once you have trained.

CAREER OPPORTUNITIES

TRAINING IN KEY PLUMBING SKILLS

CUTTING, BENDING AND JOINING OF ALL PIPE WORK
CUTTING, SHAPING AND FIXING OF SHEET ROOFING MATERIALS
DIAGNOSIS OF FAULTS IN HEATING SYSTEMS

MORE EXPERIENCE NVQ LEVEL 3 UPWARDS

FINISH APPRENTICESHIPS IN ALL PLUMBING SKILLS AREAS

FURTHER EXPERIENCE NVQ LEVEL 3/4 TRAINING FOR MANAGEMENT

SUPERVISOR/TEAM LEADER

FURTHER EXPERIENCE NVQ LEVEL 4 & 5

FOREMAN/SITE MANAGER ✦ PLANNING
MANAGEMENT ✦ TEACHING

choose to expand. The best thing to do is to talk to someone you know who has set up their own business, take their advice about the potential benefits and pitfalls, and make your own decision about whether you want to make this one of your aims for the future.

There are also plenty of career openings for plumbers to work as **lecturers** and train other plumbers in various colleges of further education and training centres. For example, Tony Beaumont was a CORGI registered plumber by trade for 25 years, working in both the commercial and industrial markets, before becoming a lecturer at South Thames College. The only danger here is that you might start to get a little detached from all the new advances that are hitting the industry. One way around this is to combine your lecturing with continuing to work as a jobbing plumber. Another is to attend the various technical evenings and seminars run by the IPHE to keep its members up to date with developments. In fact, becoming a member of the IPHE is a very good way to further your career. Again, listen to what lecturers you know have to say, and see whether this step might be right for you.

If you're working for one of the medium to large businesses, they will often present you with the opportunity for advancement into **technical**, **supervisory** or **managerial** jobs. These may include the supervision of installation work, business management, maybe the design of complex systems or providing project estimates. If you are clear that this is what you'd prefer to be doing rather than just straightforward working with the tools, then by all means explore the option. The main point here is that you will probably have the chance to watch your own supervisors or managers at work so you can see for yourself what sort of problems they have to deal with in their working lives. If the thought of following them inspires you, you've given yourself a good start towards achieving this goal.

As was mentioned earlier on in this book, there should be no shortage of jobs on the market if you are really determined to find one. However there is no guarantee that the vacancies will be evenly distributed throughout the country so it's best to check with your local careers office before you start applying.

Careers advisers can put you in touch with plumbing companies in your area, or you can look them up in the *Yellow Pages* and write to them yourself. You can also use internet search engines to look for plumbers or heating engineers.

Alternatively, if you want to find approved qualified plumbers, you can either go to the IPHE website or check out your local trade directory. Remember to keep a note of the progress of your enquiries. It's best to be thorough and methodical in your search. For instance, it might be a good idea to start by contacting prospects in your local town or area. If that draws a blank, then spread the net to bring in adjacent towns and areas – and go on

DID YOU KNOW?

City high-flyer, 33-year-old Matthew Brumwell gave up a six-figure salary and Porsche in order to start a new life as a plumber. He attended Lambeth College in London, doing evening classes and qualifying to become a CORGI-registered gas installer. He set up Cotswold Plumbers in Cheltenham with his wife Deborah and now earns about £40,000 a year. 'This wasn't about the money' he says, 'this was about my sanity and happiness.'

broadening your approach until you get a favourable response.

The important point is that you must not get downcast by people turning you away. It's a highly competitive job market and employers get many applications for each position so you must not take any rejections personally. Eventually your persistence will pay off.

PREPARING YOUR CV

It's a good idea to review constantly the way that your message is coming across to possible employers. Your character and personality cannot register with them until they've seen you. Bear in mind you've got to convince people to grant you an interview in the first place. Here are a few suggestions on presenting yourself on paper to potential employers by writing an eye-catching CV and a punchy covering letter.

Remember what the function of a CV is, namely to make you stand out from the crowd and get you an interview. So keep your CV short, no more than two pages, but provide employers with the sort of detail that will bring you to their notice.

If you haven't yet had many, or even any, jobs – don't worry. The sort of details that it will be useful to concentrate on are your achievements and the skills you've acquired – and make sure that those skills correspond with those that any reputable company will be looking for.

Lay out your CV neatly and clearly, whether it is being sent by post or email. Check carefully to make sure everything reads well and is free from mistakes. Compose your covering letter with great care so that it conveys your enthusiasm for joining the company without indulging in overselling. For a full examination of how to present yourself on paper, try *Winning CVs for First-time Job Hunters* by Kathleen Houston (Trotman), which has excellent advice on every stage of the process.

Here is a list of the most important things you can do to make sure you are the candidate a prospective employer will be willing to take on:

- Do your research on your target company – in depth.
- Prepare a relevant, customised CV.
- Show you've got the energy, commitment and aptitude to be a plumber.
- Gain any work experience you can.
- Read everything about the industry you can get your hands on.
- Apply for training programmes at your natural level (e.g. Apprenticeships).
- Snap up any free courses.
- Develop other related key skills.
- Be ready to do anything to gain a start.
- Take all the luck you can find (but remember the harder you work, the luckier you get).

For more information on job-hunting and interviews see *Winning Job-hunting Strategies for First-time Job Hunters* by Gary Woodward (Trotman) and *Winning Interviews for First-time Job Hunters* by Kathleen Houston (Trotman).

If you have made it this far through the book then you should know if **Plumbing** really is the career for you. But, before contacting the professional bodies listed in the next chapter, here's a final, fun checklist to show if you have chosen wisely.

THE LAST WORD   ✔ TICK YES OR NO

DO YOU LIKE WORKING WITH YOUR HANDS?
☐ YES
☐ NO

DO YOU LIKE WORKING WITH PEOPLE?
☐ YES
☐ NO

DO YOU WANT A JOB WHERE YOU WILL BE DOING SOMETHING DIFFERENT EVERY DAY?
☐ YES
☐ NO

ARE YOU SELF-MOTIVATED AND ABLE TO THINK ON YOUR FEET?
☐ YES
☐ NO

ARE YOU ABLE TO COMMUNICATE EFFECTIVELY WITH LOTS OF DIFFERENT PEOPLE?
☐ YES
☐ NO

ARE YOU A SELF-STARTER, ABLE TO TAKE CONTROL AND RESPONSIBILITY?
☐ YES
☐ NO

If you answered 'YES' to all these questions then
CONGRATULATIONS! YOU'VE CHOSEN THE RIGHT CAREER

If you answered 'NO' to any of these questions then this may not be the career for you.
However, there are still some options open to you,
for example you could work as a Sales person at a Plumbers' Merchant or DIY store.

12

Resources

In this section you will find all the addresses, telephone numbers and websites for the relevant government and industry advisory and training bodies for the plumbing industry. There is also a list of publications you may find useful to read.

TRAINING AND ADVICE

British Plumbing Employers' Council (Training) Ltd (BPEC)
2 Mallard Way
Pride Park
Derby DE24 8GX
T: 0845 644 6558
www.bpec.org.uk

BPEC operates as a charitable trust funding appropriate training-related initiatives for the plumbing industry. Its website gives some great information on training opportunities.

City and Guilds
1 Giltspur Street
London EC1A 9DD
T: 020 7294 2800
www.cityandguilds.com

City and Guilds is the leading provider of vocational qualifications in the UK. It has five different levels of qualification, with Level 1 based on the lowest competence level and Level 5 based on the highest, and it offers everything from NVQ and SVQ to Apprenticeships and Higher Level Qualifications. The website lists all the qualifications it provides in plumbing.

Connexions
www.connexions-direct.com

The Connexions service has been set up especially for young people aged 13 to 19 and offers advice, support and practical help on many subjects, including future career options. On the Connexions website you will find the jobs4u careers database. Search this database for information on training and jobs in plumbing.

CITB-ConstructionSkills
Bircham Newton
King's Lynn
Norfolk PE31 6RH
T: 01485 577577
www.citb-constructionskills.co.uk

CITB-ConstructionSkills is committed to training professionals to a very high level. It has some very good pamphlets and brochures offering more information; alternatively have a look at the website.

Council for Registered Gas Installers (CORGI)
1 Elmwood
Chineham Park
Crockford Lane
Basingstoke RG24 8WG

T: 0870 401 2200
www.trustcorgi.com

This is the national watchdog for gas safety in the UK and registration is a legal requirement for both businesses and self-employed people working on gas fittings, so if you want to work in this area once you are qualified you will have to register. Registration details can be found on the website in the Members section.

Department for Education and Skills (DfES)
Packs available from 0800 585505
www.dfes.gov.uk

If you are undertaking a vocational training course lasting up to two years (with one year's practical work experience if it is part of the course) you may be eligible for a Career Development Loan (see www.lifelonglearning.co.uk for more info). These are available for full-time, part-time and distance learning courses and applicants can be employed, self-employed or unemployed. The DfES pays the interest on the loan for the length of the course and up to one month afterwards. You can also find out which colleges are Centres of Vocational Excellence for plumbing on the website (www.dfes.gov.uk/coves).

Edexcel
One90 High Holborn
London WC1 7BB
T: 0870 240 9800
www.edexcel.org.uk

Edexcel has taken over from BTEC in offering BTEC qualifications including BTEC First Diplomas, BTEC National

Diplomas and BTEC Higher Nationals (HND and HNC). It also offers NVQ qualifications. The website includes qualification 'quick links' and you can search by the qualification or the career you are interested in.

Institute of Domestic Heating Engineers and Environmental Engineers (IDHEE)

35 New Forest Enterprises Centre
Chapel Lane
Totton
Southampton SO40 9LA
T: 023 8066 8900
www.idhee.org.uk

IDHE is an independent, non-profit-making professional body founded in 1964. It is solely for the domestic heating industry.

Institute of Plumbing and Heating Engineering (IPHE)

64 Station Lane
Hornchurch
Essex RM12 6NB
T: 01708 472791
www.iphe.org.uk

Founded in 1906, this is the UK's professional body for plumbers and others in the industry and has over 12,400 members. A registered educational charity, it is also secretariat to the World Plumbing Council. The website is visited by over 1,300 people a day and has a very good careers section and links to other related bodies as well as a very interesting section on women in plumbing.

SORT IT OUT!

HOW DO I KNOW WHICH JOBS ARE RIGHT FOR ME?

No problem, you can log onto **cityandguilds.com/myperfectjob** and take 20 minutes to answer a range of online questions which looks at your interests, personality and lifestyle and suggests job areas which may suit you. Get all the information on job options, how to get started and where you can go to study.

cityandguilds.com/myperfectjob

Learning and Skills Council
Apprenticeship Helpline: 0800 015 0600
www.lsc.gov.uk
www.apprenticeships.org.uk

Launched in 2001, Learning and Skills Council now
has a main office in Coventry and nine regional offices.
It is responsible for the largest investment in post-16
education and training in England and this includes
further education colleges, work-based training and
workforce developments. For 2006/7 its budget is
£10.4 billion.

For Modern Apprenticeships in Scotland:
www.modernapprenticeships.com
www.careers.scotland.org.uk

In Wales:
www.beskilled.net

New Deal
T: 0845 606 2626
www.newdeal.co.uk

If you are an older individual looking to change careers
and you have been unemployed for six months or more
(or receiving Jobseekers Allowance), you may be able to
gain access to NVQ/SVQ courses through the New Deal
programme. People with disabilities, ex-offenders and
lone parents are eligible before reaching six months of
unemployment. Check out the website for more
information.

Plumbing and Heating Industry Alliance (PHIA)
www.phia.org.uk

Born out of the Plumbing Forum, PHIA has 19 members,
including the Worshipful Company of Plumbers and the
Institute of Domestic Heating Engineers and Environmental
Engineers. The website contains contact details for each of
its members.

Qualifications and Curriculum Authority (QCA)
83 Piccadilly
London W1J 8QA
T: 020 7509 5555
www.qca.org.uk

In Scotland:
Scottish Qualifications Authority (SQA)
Optima Building
58 Robertson Street
Glasgow G2 8DQ
Customer Contact Centre: 0141 242 2214
www.sqa.org.uk

These official awarding bodies will be able to tell you whether
the course you choose leads to a nationally approved
qualification such as an NVQ or SVQ.

**Scotland and Northern Ireland Plumbing Employers'
Federation (SNIPEF)**
2 Walker Street
Edinburgh EH3 7LB
T: 0131 225 2255
www.snipef.org.uk
www.becomeaplumber.org

Founded in 1923, SNIPEF is the trade association for businesses involved in installing and maintaining plumbing and heating systems throughout Scotland and Northern Ireland.

SummitSkills
Vega House
Opal Drive
Fox Milne
Milton Keynes MK15 0DF
T: 01908 303960
www.summitskills.org.uk

This is the Sector Skills Council for the electrotechnical, heating, ventilating, air conditioning, refrigeration and plumbing industries. It aims to champion the training and development of all these industries. You can find out which plumbing courses are accredited through SummitSkills.

PUBLICATIONS

Heating and Plumbing Monthly
Becket House
Vestry Road
Sevenoaks TN14 5EJ
T: 01732 748000
www.unity-media.com

A monthly magazine aimed at heating engineers and plumbers working in the domestic and light commercial sectors.

Plumb Heat
2 Walker Street
Edinburgh EH3 7LB
T: 0131 225 2255
www.snipef.org.uk

A quarterly journal containing information for the Scotland and Northern Ireland Plumbing Employers' Federation.

Plumbing and Heating in Northern Ireland
13B Lisburn Road
Moira
Craigavon BT63 0JR
T: 028 9261 2990
www.plumbingandheatingmagazine.co.uk

This quarterly journal contains industry and product news for the Northern Irish market.

P&HE
64 Station Lane
Hornchurch RM12 6NB
T: 01708 463114
www.iphe.org.uk

This is the bi-monthly journal of the Institute of Plumbing and Heating Engineers giving industry news aimed at members.